STAMPABILITY

****

STARS

STEWART & SALLY WALTON

PHOTOGRAPHY BY GRAHAM RAE

LORENZ BOOKS

NEW YORK · LONDON · SYDNEY · BATH

CONTENTS

*I*NTRODUCTION

EVERY NOW AND THEN there is a breakthrough in interior decorating –
something suddenly captures the imagination. Stamping is definitely
one such breakthrough and it is all the more popular as it needs neither
special knowledge nor lots of money.

All you need is a stamp and some color and you're ready. The idea
comes from the office rubber stamp and it uses the same principle. You
can use stamps with a stamp pad, but a small foam roller gives a better
effect. The stamp can be coated with ordinary household paint – this
makes stamping a fairly inexpensive option, and gives you a wide range
of colors to choose from.

There are stamping projects in this book ranging from a single stamped
print on a cushion to the bathroom of your dreams. Each project is
illustrated with clear step-by-step photographs and instructions. You are
bound to progress onto your own projects once you've tried these
suggestions because stamping really is easy. The added bonus is that you
need very little equipment and there's hardly any clearing up to do
afterward – what could be better?

This book focuses on the star theme. Stars have been used decoratively
in many cultures throughout history. The shape may have spiky or
chunky points but the star image remains recognizable and powerful
wherever it is used. The great thing about star designs is that they never
go out of fashion and they can be interpreted in a wide variety
of styles and colors.

So, whether your tastes lean toward dramatic dark backgrounds with
shimmering stars, the romance of a country cottage, or the sheer
elegance of gold on white, you are sure to find something to inspire you
in this collection.

Basic Application Techniques

Stamping is a simple and direct way of making a print. The variations, such as they are, come from the way in which the stamp is inked and the type of surface to which it is applied. The stamps used in the projects were inked with a foam roller which is easy to do and gives reliable results, but each application technique has its own character. It is a good idea to experiment and find the method and effect that you most prefer.

INKING WITH A BRUSH

The advantage of this technique is that you can see where the color has been applied. This method is quite time-consuming, so use it for smaller projects. It is ideal for inking an intricate stamp with more than one color.

INKING WITH A FOAM ROLLER

This is the best method for stamping large areas, such as walls. The stamp is evenly inked and you can see where the color has been applied. Variations in the print effects can be achieved by re-inking the stamp after several printings.

INKING ON A STAMP PAD

This is the traditional way to ink rubber stamps, which are less porous than foam stamps. The method suits small projects, particularly printing on paper. Stamp pads are more expensive to use than paint but are less messy and produce very crisp prints.

INKING BY DIPPING IN PAINT

Spread a thin layer of paint onto a flat plate and dip the stamp into it. This is the quickest way of stamping large decorating projects. As you cannot see how much paint the stamp is picking up, you will need to experiment.

INKING WITH FABRIC PAINT

Spread a thin layer of fabric paint on a flat plate and dip the stamp into it. Fabric paints are quite sticky and any excess paint is likely to be taken up in the fabric rather than to spread. Fabric paint can also be applied by brush or foam roller and is available with built-in applicators.

INKING WITH SEVERAL COLORS

A brush is the preferred option when using more than one color on a stamp. It allows greater accuracy than a foam roller because you can see exactly where you are putting the color. Two-color stamping is very effective for giving a shadow effect or a decorative pattern.

SURFACE APPLICATIONS

The surface onto which you stamp your design will greatly influence the finished effect.
Below are just some of the effects that can be achieved.

STAMPING ON ROUGH PLASTER

You can roughen your walls before stamping by mixing joint compound to a fairly loose consistency and spreading it randomly on the wall. When dry, roughen with coarse sandpaper, using random strokes.

STAMPING ON SMOOTH PLASTER OR WALLPAPER LINER

Ink the stamp with a small foam roller for the crispest print. You can create perfect repeats by re-inking with every print, whereas making several prints between inkings varies the strength of the prints and is more in keeping with hand-printing.

STAMPING ON WOOD

Rub down the surface of any wood to give the paint a better "tooth" to adhere to. Some woods are very porous and absorb paint, but you can intensify the color by over-printing later. Wood looks best stamp lightly so that the grain shows through. Seal your design with clear matte varnish

STAMPING ON GLASS

Wash glass in hot water and detergent to remove any dirt or grease and dry thoroughly. It is best to stamp on glass for non-food uses, such as vases or sun-catchers. Ink the stamp with a foam roller and practice on a spare sheet of glass. As glass has a slippery, non-porous surface, you need to apply the stamp with a direct on/off movement. Each print will have a slightly different character, and the glass's transparency allows the pattern to be viewed from all sides.

STAMPING ON TILES

Wash and dry glazed tiles thoroughly before stamping. If the tiles are already on the wall, avoid stamping in areas that require a lot of cleaning. The paint will only withstand a gentle wipe with a damp cloth. Loose tiles can be baked to add strength and permanence to the paint. Read the paint manufacturer's instructions (and disclaimers!) before you do this. Ink the stamp with a small foam roller and apply with a direct on/off movement.

STAMPING ON FABRIC

As a rule, natural fabrics are the most absorbent, but to judge the stamped effect experiment on a small sample. Fabric paints come in a range of colors, but to obtain the subtler shades you may need to combine the primaries and black and white. Always place a sheet of cardboard behind your work surface. Apply the fabric paint with a foam roller, brush or by dipping. You will need more paint than for a wall, as fabric absorbs the paint more efficiently.

$\mathcal{P}$AINT EFFECTS

Once you have mastered the basics of stamp decorating, there are other techniques that you can use to enrich the patterns and add variety. Stamped patterns can be glazed over, rubbed back or over-printed to inject subtle or dramatic character changes.

STAMPING LATEX PAINT ON PLASTER, DISTRESSED WITH TINTED VARNISH

The stamped pattern will already have picked up the irregularities of the wall surface and, if you re-ink after several prints, some prints will look more faded than others. To give the appearance of old hand-blocked wallpaper, paint over the whole surface with a ready-mixed antiquing varnish. You can also add color to a clear varnish, but never mix a water-based product with an oil-based one.

STAMPING LATEX PAINT ON PLASTER, COLORED WITH TINTED VARNISH

If the stamped prints have dried to a brighter or duller shade than you had hoped for you can apply a coat of colored varnish. It is possible to buy ready-mixed color-tinted varnish, or you can add color to a clear varnish base. A blue tint will change a red into purple, a red will change yellow into orange, and so on. The color changes are gentle because the background changes at the same time.

STAMPING WITH WALLPAPER PASTE, WHITE GLUE AND WATERCOLOR PAINT

Mix three parts pre-mixed wallpaper paste with one part white glue and add watercolors. These come ready-mixed in bottles with built-in droppers. The colors are intense, so you may only need a few drops. The combination gives a sticky substance which the stamp picks up well and which clings to the wall without drips. The white glue dries clear to give a bright, glazed finish.

STAMPING WITH A MIXTURE OF WALLPAPER PASTE AND LATEX PAINT

Mix up some wallpaper paste and add one part to two parts latex paint. This mixture makes a thicker print that is less opaque than the usual latex version. It also has a glazed surface that picks up the light.

STAMPING LATEX PAINT ON PLASTER, WITH A SHADOW EFFECT

Applying even pressure gives a flat, regular print. By pressing down more firmly on one side of the stamp, you can create a shadow effect on one edge. This is most effective if you repeat the process, placing the emphasis on the same side each time.

STAMPING A DROPPED SHADOW

To make a pattern appear three-dimensional, stamp each pattern twice. Make the first print in a dark color that shows up well against a mid-tone background. For the second print, move the stamp slightly to one side and use a lighter color.

DESIGNING WITH STAMPS

To design the stamp pattern, you need to find a compromise between printing totally at random, and measuring precisely to achieve a machine-printed regularity. You can use the stamp block itself to give you a means of measuring your pattern, or try strips of paper, squares of cardboard and lengths of string. Try using a stamp pad on scrap paper to plan your design but always wash and dry the stamp before proceeding to the main event.

1 USING PAPER CUT-OUTS
The easiest way to plan your design is to stamp and cut out as many pattern elements as you need and use them to mark the position of your finished stamped prints.

2 CREATING A REPEAT PATTERN
Use a strip of paper as a measuring device for repeat patterns. Cut the strip the length of one row of the pattern. Use the stamp block to mark where each print will go, with equal spaces in between.

3 USING A PAPER SPACING DEVICE
This method is very simple. Decide on the distance between prints and cut a strip of paper to that size. Each time you stamp, place the strip against the edge of the previous print and line up the edge of the stamp with the other side of the strip. Use a longer strip to measure the distance between rows.

4 CREATING IRREGULAR PATTERNS
If your design does not fit into a regular grid, practice the pattern first on paper. Cut out paper shapes and use these to position the finished pattern. Alternatively, raise a motif above the previous one by stamping above a strip of cardboard positioned on the baseline.

5 DEVISING A LARGER MOTIF
Use the stamps in groups to make up a larger design. Try stamping four together in a a block, or partially overlapping an edge so that only a section of the stamp is shown. Use the stamps upside down, back to back and rotated in different ways. Experiment on scrap paper first.

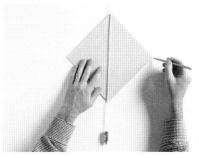

6 USING A PLUMBLINE
Attach the plumbline at ceiling height, to hang down the wall. Hold a cardboard square behind the plumbline so that the string cuts through two opposite corners. Mark all four points, then move the cardboard square down. Continue in this way to make a grid.

QUILTED THROW PILLOWS

A new pile of throw pillows can change the mood of a room in an instant – so why not update your living room with these folk-inspired patterns? You can vary the star design on each pillow to make a coordinating mix-and-match set. Don't be daunted by the idea of quilting, as it really is very easy. Simply iron the backing, called quilter's batting, onto the calico, then sew straight, horizontal, vertical or diagonal lines through the centers of the stars. Alternatively, buy ready-made covers and stamp on the star pattern.

YOU WILL NEED
19 x 19in unbleached calico (washed and ironed)
backing paper (such as thin cardboard or newspaper)
fabric paint in brick-red and pear-green
plate
foam roller
folk-art and small star stamps
iron
iron-on quilter's batting
needle and thread
19 x 19in plain-colored fabric

1 Lay the calico on some backing paper on a work surface. Spread some of the brick-red paint onto the plate and run the roller through it until it is evenly coated. Ink the folk-art stamp.

2 Position the stamp in the first corner of the calico and print.

3 Continue stamping along the row, leaving a space the width of the stamp block between each print. Begin the next row with a blank space, and stamp the star so that it falls between the stars in the bottom row. Repeat these two rows to cover the fabric.

4 Using the pear-green paint, ink the small star stamp and make a print in each space between the brick-red stars. Fix the fabric paint with a hot iron following the manufacturer's instructions and iron on the batting. To make the quilted pattern, sew horizontal and then vertical lines through the centers of the red stars.

STARRY BEDROOM

At first glance, this bedroom looks wallpapered, but a closer examination reveals the hand-printed irregularity of the star stamps – some are almost solid color while others look very faded. This effect is achieved by making several prints before re-inking the stamp. The idea is to get away from the monotony of machine-printed wallpaper, where one motif is the exact replica of the next, and create the effect of exclusive, hand-blocked wallpaper at a fraction of the price. The grid for the stars is marked using a plumbline and pencil. If you haven't got a plumbline, make your own by tying a key to a piece of string. You're bound to be delighted with the final result, and feel a great sense of achievement at having done it all yourself.

YOU WILL NEED
sandy-yellow latex or distemper paint
paintbrush
12 x 12in cardboard
plumbline
pencil
brick-red latex paint (matchpot size)
plate
foam roller
folk-art star stamp

1 Paint the walls sandy yellow with latex or distemper. You may prefer a smooth, even finish or areas of patchy color – each will create its own distinct look.

2 Hold the cardboard square diagonally against the wall in the corner at ceiling height. Attach the plumbline at ceiling height so that the string cuts through the top and bottom corners of the square. Mark all four points in pencil. Continue moving the square and marking points to form a grid.

3 Spread some brick-red paint onto the plate and run the roller through it until it is evenly coated. Ink the stamp and print a star on every pencil mark, or line the block up against each pencil mark to find your position, whichever you find easiest.

4 Experiment with the stamp and paint to see how many prints you can make before re-inking. Don't make the contrast between the pale and dark too obvious or the eye will always be drawn to these areas.

Small Cabinet

A popular designer's trick is to paint a piece of furniture in the same colors as the background of the room, but in reverse. This coordinates the room without being overpoweringly repetitive. A small cabinet like the one in this project is perfect for such a treatment. Don't be too precise in your stamping – a fairly rough-and-ready technique gives the most pleasing results.

YOU WILL NEED
wooden cabinet
latex paint in brick-red and yellow
paintbrush
plate
foam roller
small star stamp
fine steel wool or sandpaper
water-based matte varnish and brush

1 Paint the cabinet with a base coat of brick-red latex.

2 Spread some yellow paint onto the plate and run the roller through it until it is evenly coated. Ink the stamp and print onto the cabinet.

3 Rub around the edges with steel wool or sandpaper to simulate natural wear and tear. This will give an aged appearance sympathetic to a country-style interior.

4 Apply a coat of varnish tinted with the brick-red paint (one part paint to five parts varnish) to tone down the contrast between the prints and the background paint.

HEADBOARD

This wooden headboard was the ideal candidate to be given a touch of individuality with a single star pattern in a single color. The stamped star pattern could not be easier to apply. Begin stamping in one corner and follow the shape of the headboard, judging the spacing by eye. The design can be adapted to make a fabulous circus-style bed for a child, simply by changing the colors. Try a bright yellow background with scarlet stars or silver with red and blue for a dazzling contrast.

YOU WILL NEED
wooden headboard
latex paint in red-brown, forest-green and
light blue
paintbrush
plate
foam roller
small star stamp
water-based matte varnish and brush

1 Paint the headboard in red-brown and forest-green and let dry. Spread a small amount of light blue paint onto the plate and run the roller through it until it is evenly coated.

2 Ink the stamp and begin printing along the top of the headboard.

3 Stamp the stars down the corner posts on the flat surfaces between the turned sections. Apply a coat of varnish to protect the surface.

GLASS VASE

The transparency of glass gives a new dimension to the stamped stars. The color is applied to one surface, but the design is visible from all sides. You could try the stamps on any plain glass vase – this one was particularly easy to work with because of the flat surfaces.

There are now some paints available called acrylic enamels. These are suitable for use on glass and ceramics and they give a hard-wearing finish that stands up to non-abrasive washing. The selection of colors is great, so take a look at them and try some glass stamping.

YOU WILL NEED
glass vase
dish towel
dark-colored acrylic enamel paint
plate
foam roller
large star stamp
piece of glass

1 Wash the vase to remove any grease from the surface. Dry it thoroughly.

2 Spread some paint onto the plate and run the roller through it until it is evenly coated. Ink the stamp and make a test print on the piece of glass.

3 Stamp the stars onto the glass vase. Apply gentle pressure with a steady hand and remove the stamp directly to avoid it sliding on the slippery surface.

LAMP AND LAMPSHADE

Transforming mass-produced accessories like this lamp is almost as satisfying as making them from scratch. Bases and shades can be very dull, but with some stamp decoration they soon become cheerful and individualized. Choose a gentle color, like a light blue, and when the paint has dried, rub it down to the wood in places with fine sandpaper or steel wool. Try not to be too heavy-handed – a subtle, aged look is most effective. When you stamp the lampshade, use the same light blue paint for most of the stars and add a few red ones as highlights. It is a good idea to stamp some stars partially overlapping the lampshade edge to give a good all-over pattern.

YOU WILL NEED
wooden lamp base
light blue latex paint
paintbrush
fine steel wool or sandpaper
plate
foam roller
small star stamp
plain cream fabric lampshade
brick-red latex or fabric paint (only a very small amount is needed, so use what is available)

1 Apply a coat of light blue latex to the lamp base and let dry thoroughly.

2 Using either fine steel wool or sandpaper, rub the base down to bare wood along the moldings.

3 Spread some light blue paint onto the plate and run the roller through it until it is evenly coated. Ink the stamp and print the stars onto the shade, leaving more or less equal spaces between them with a few larger spaces for the brick-red stars – use the stamp block to judge the size of the spaces.

4 Stamp brick-red stars in the spaces. Partially over-print the edges of the shade in some places.

WRAPPING PAPER

Imagine never having to buy a sheet of wrapping paper again. Once you realize how easy it is to make stamped paper, there will be no turning back. Stamped paper not only looks great, but it costs so little that you will have lots of spare cash to spend on ribbon and trimmings.

Any plain paper can be stamped. The contrast between utilitarian brown wrapping paper and luxurious gold stars looks particularly striking, but for a more colorful paper, you could use tissue paper, which is available in a riotous assortment of colors.

Only the most exclusive and expensive types of commercial wrapping paper are hand-printed, so you can have the pleasure of stamping your own at a fraction of the cost.

YOU WILL NEED
ruler
pencil
brown wrapping paper
acrylic paint in brown, blue, white
and cream
plate
foam roller
starburst, folk-art and small star stamps

1 Use the ruler and pencil to mark one edge of the wrapping paper at approximately 5in intervals.

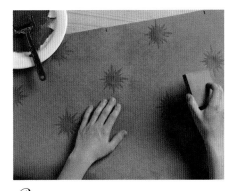

2 Spread a small amount of brown paint onto the plate and run the roller through it until it is evenly coated. Ink the starburst stamp and print onto the wrapping paper, using the pencil marks as a guide for the first row and judging the next rows visually.

3 Ink the folk-art stamp with blue paint and stamp these stars between the brown stars.

4 Ink the small star stamp with white paint and fill in the centers of the blue stars.

5 Stamp cream stars along diagonal lines between the blue stars.

$\mathscr{S}$TAR-STUDDED BATHROOM

Transform your bathroom into one fit for royalty or Hollywood stars in the space of a weekend.
The cost is minimal and the luxurious result is bound to impress all future bathers.
If you have an old bathroom with a boxed-in tub, it is worth investigating the underside – you
may be lucky enough to discover an old claw-footed iron tub like this one. However, the lack of
an elegant tub just adds to the challenge. You can paint the boarding and arrange it with drapes to
soften the edges.
Who could fail to feel pampered in such wonderful surroundings?

YOU WILL NEED

tape measure
small semicircular wooden shelf
pelform (see step 1)
craft knife
ruler
pencil
piece of calico (see step 3)
gold fabric paint or white glue mixed with
bronze powder
plate
foam roller
large star stamp
gold tassels
white glue or needle and matching
sewing threads
length of cheesecloth, for the curtains
iron-on hem tape
iron-on header tape
iron
backing paper (such as thin cardboard
or newspaper)
staple gun and staples
thumbtacks
round satin cushion
metal primer (optional)
paintbrush
cream enamel or latex paint

1 To make the valance, measure the curved edge of the shelf and cut the pelform to this length.

2 Using a ruler and pencil, draw a series of points along the bottom of the pelform. Cut them out.

3 Place the pelform over a rectangle of calico and cut around the shape, leaving a 1¼in seam allowance. Peel off the backing paper and smooth the fabric onto the pelform. Turn the piece over and snip into the corners. Then peel off the backing strip and stick the seams down.

4 Spread some gold fabric paint onto the plate and run the roller through it until it is evenly coated. Ink the stamp and print a row of stars across the middle of the valance. Glue or sew a tassel to each point. Now fix the shelf to the wall, about 12in from the ceiling, exactly halfway along the length of the tub. ▶

5 To make the curtains, cut the cheesecloth into two pieces and turn up a seam at both ends using iron-on hem tape. Attach iron-on header tape to one end of each curtain. Lay out one of the curtains on backing paper. Spread some gold fabric paint onto the plate and run the roller through it until it is evenly coated. Ink the stamp and print a widely spaced row of stars across the bottom of the fabric.

6 Print another row of stars to fall between the stars in the first row. Repeat these rows to cover the curtain. Stamp the second curtain in the same way.

7 Gather up the curtains by pulling the header tape string. Staple the curtains to the shelf.

8 Attach the valance to the shelf using thumbtacks.

9 To make the cushion, place a ruler diagonally across the cushion and make a small pencil mark in the center. Check the position by intersecting the center point from a different angle.

10 Spread some gold fabric paint on to the plate and run the roller through it until it is evenly coated. Ink the stamp and print a single star in the center of the cushion.

11 Paint the tub with a coat of metal primer, if necessary. Paint the tub or tub panel cream and let dry. Paint the legs gold. Spread some gold fabric paint on to the plate and run the roller through it until it is evenly coated. Ink the large stamp and begin by printing the top row of stars. Continue stamping in the same way as for the curtains so that the stars fall between the spaces of the row above. Follow the curve of the tub.

*F*LOORCLOTH

A floorcloth is a sheet of canvas, painted and varnished and used in place of a rug. Canvas is hardwearing, especially after a few coats of varnish, and it feels cool and smooth underfoot. Artist's canvas comes in all widths – just think of the paintings you've seen – and it is available through arts and crafts suppliers. Floorcloths were originally used by American settlers who found that sailcloth could be stretched over a bed of straw to cover hard floors. They decorated them to imitate checkered marble floors and fine carpets, and found that the paint and varnish added to their durability. They were eventually replaced by linoleum and other cheap floor coverings, but have once again become popular in country-style interiors.

YOU WILL NEED
cream artist's canvas
pencil
ruler
scissors
fabric glue and brush
white acrylic primer
paintbrush
latex paint in dark blue, lime-green and light blue
plate
foam roller
starburst stamp
matte varnish and brush

1 Draw a 1½in border around the edge of the canvas, then fold this back to make a seam. Miter the corners by cutting across them at a 45-degree angle, then apply the glue and paste down the edges. Prepare the canvas with white acrylic primer.

2 Paint the primed floorcloth dark blue, applying two coats if necessary. Measure and draw a 4in border around the edge.

3 Paint the border lime-green, applying two coats for good coverage.

4 Spread some light blue paint onto the plate and run the roller through it until it is evenly coated. Ink the stamp and begin printing the stars in one corner. Judge the spacing visually, stamping a random arrangement of stars to cover the cloth.

5 Apply several coats of hard-wearing matte varnish, letting each one dry thoroughly before applying the next.

Striped Foyer

All too often, creative decorating is restricted to the larger rooms of a house, but the foyer is the first thing everybody sees when they come through the front door. Why not use stamping to make a stunning first impression on your visitors?

The color scheme combines two earth colors with bright silver stars to give a slightly Moroccan feel. If you have an immovable carpet or tiles that don't suit these shades, then choose a color from your existing floor covering to highlight the walls. Foyers seldom have windows to give natural light and the inclination is to use light, bright colors to prevent them from looking gloomy. A better idea is to go for intense, dramatic colors with good electric lighting – they will turn a corridor into a welcoming foyer.

YOU WILL NEED
household sponge
latex paint in light coffee and
spicy brown
pencil
plumbline
reusable adhesive
straight-edged cardboard or wooden board,
the width required for the stamps
paintbrush
scissors
silver acrylic paint
plate
foam roller
starburst stamp

1 Use a household sponge to apply irregular patches of light coffee latex paint to the wall. Let dry.

2 Attach a plumbline at ceiling height with reusable adhesive so it hangs just away from the wall. Line up one straight edge of the cardboard or wooden board and use this as a guide to draw a straight line in pencil down the wall.

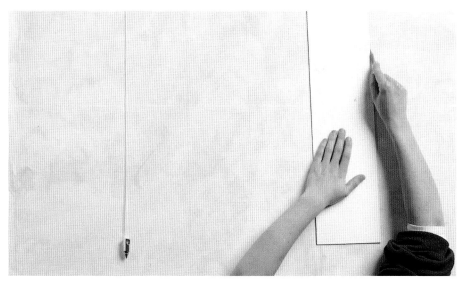

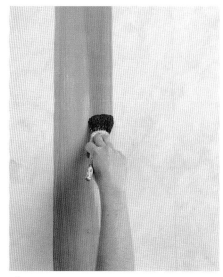

3 Move the cardboard marker one width space along the wall, and continue to mark evenly spaced lines.

4 Paint the first stripe spicy brown. Try to keep within the pencil lines, but don't worry too much about slight mistakes as the wall should look hand-painted and not have the total regularity of wallpaper.

5 Continue painting each alternate stripe, keeping within the pencil lines, but attempting to create a slightly irregular finish.

6 Cut the cardboard spacer to the length required to use as a positional guide for the stars. Spread some silver paint onto the plate and run the roller through it until it is evenly coated. Ink the stamp and print a star in each of

the spicy brown stripes along the wall, above and below the spacer. Continue printing across the wall, then return to the first stripe and start printing again, one space below the lowest star. ▷

7 Using the spacer as before, print the first two rows of stars in the coffee-colored stripes. Position these stars so they fall midway between the stars in the spicy brown stripe.

8 Continue this process to fill in the remaining stars all down thecoffee-colored stripes.

PICTURE FRAME

This project combines all the creative possibilities of stamping. It involves four processes: painting a background, stamping in one color, over-printing in a second color and rubbing down to the wood. These processes transform a plain wooden frame and they are neither time-consuming nor expensive.

It is surprisingly difficult to find small, old frames that are broad enough to stamp. Fortunately, a wide range of basic, cheap frames can be found in craft stores.

YOU WILL NEED
picture frame
latex paint in sky-blue, red-brown
and gold
paintbrush
plate
foam roller
small and large star stamp
fine steel wool or sandpaper

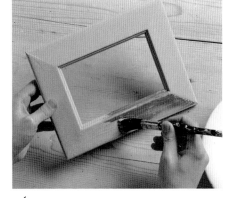

1 Paint the frame sky-blue and let dry thoroughly.

2 Spread a small amount of red-brown paint onto the plate and run the roller through it until it is evenly coated. Ink the first stamp and print it in the middle of each side.

3 Using the red-brown paint, stamp a large star over each corner. Let dry thoroughly.

4 Ink the large stamp with gold and over-print the red-brown corner stars. Let dry before rubbing the frame gently with steel wool or sandpaper. Experiment with dropped shadow effects and other designs.

This edition published in 1996 by Lorenz Books
an imprint of Anness Publishing Limited
Administrative office:
27 West 20th Street
New York, NY 10011

© 1996 Anness Publishing Limited

Lorenz Books are available for bulk purchase, for sales promotion, and for premium use. For details write or call the manager of special sales, 27 West 20th Street, New York, NY10011; (212) 807 6739

ISBN 1 85967 230 2

Publisher: Joanna Lorenz
Senior Editor: Lindsay Porter
Assistant Editor: Sarah Ainley
Designer: Bobbie Colgate Stone
Photographer: Graham Rae
Stylist: Diana Civil

3 5 7 9 10 8 6 4 2

Printed and bound in Singapore